Renan Luigi Cavalmoretti Marcelos
Paulo R M Valentin
David L P Leite

Geography and Education

Renan Luigi Cavalmoretti Marcelos
Paulo R M Valentin
David L P Leite

Geography and Education

Academic Contributions

ScienciaScripts

Imprint
Any brand names and product names mentioned in this book are subject to trademark, brand or patent protection and are trademarks or registered trademarks of their respective holders. The use of brand names, product names, common names, trade names, product descriptions etc. even without a particular marking in this work is in no way to be construed to mean that such names may be regarded as unrestricted in respect of trademark and brand protection legislation and could thus be used by anyone.

Cover image: www.ingimage.com

This book is a translation from the original published under ISBN 978-613-9-71783-5.

Publisher:
Sciencia Scripts
is a trademark of
Dodo Books Indian Ocean Ltd. and OmniScriptum S.R.L publishing group

120 High Road, East Finchley, London, N2 9ED, United Kingdom
Str. Armeneasca 28/1, office 1, Chisinau MD-2012, Republic of Moldova, Europe
Printed at: see last page
ISBN: 978-620-7-66947-9

CONTENTS

1. ENVIRONMENTAL EDUCATION: IN DEFENCE OF CRITICAL SOCIO-ENVIRONMENTAL AWARENESS

Introduction

The main function of this work is to provide the necessary theoretical content for secondary school teachers, contributing to their ongoing training. This is theoretical research based on pedagogical approaches with a progressive bias, focussing on pedagogical practices linked to environmental education, proposing another way of thinking, planning and developing the content.

In order to achieve this goal, teaching professionals must first be open to other teaching methodologies, countering traditional and conservative pedagogical practices.

The work is based on bibliographical research aimed at achieving the proposed objectives. The literature review is qualitative in nature and was carried out through systematic reading, with summaries, abstracts and syntheses, seeking to highlight the pertinent points of each author studied.

Bibliographical research is important, according to Marconi and Lakatos (2003), because it involves all the bibliography that has already been made public in relation to the object of study, including books, magazines, scientific articles, texts in general, among others, putting the researcher in direct contact with what has already been discussed and published on the subject.

The first point revolves around the social, economic and political aspects that have guided the development of policies and approaches to environmental issues throughout history. The second point has the function of providing the necessary tools for teachers to be able to contribute to building a collective and interventionist environmental social conscience. Collaborating in the formation of people who are critical and active in relation to environmental issues.

Historical Context: "The Environmental Debate"

In order to understand the debate on human action on the environment and its multiple expressions, it is necessary to understand the historical and social context in which this discussion has been disseminated. According to SILVA (2012, p. 44) "the impact of human action on the environment varies historically according to the mode of production, class structure, technological apparatus, demographic growth and the cultural universe of each society". From this perspective, in the 20th century, the industrialised economy centred on cities and the need to search for more technology had a negative impact on the environment.

In 1962, in the United States of America, journalist Rachel Carson released her book *Silent Spring*, which was considered a milestone in the global environmental movement. The book exposed the environmental impacts generated by the excessive use of chemical products, which sparked a national debate on the responsibility of science and the limits of technological progress, initiating a transformation in the relationship between human beings and the environment.

In 1972, the Stockholm Conference was held with the aim of raising society's awareness of how to change its relationship with the environment and thus fulfil its active and present role in environmental issues that affect the lives of future generations.

This concern has also arisen due to the overuse of natural resources to supply the growing number of consumers looking for new products and services. This has led to the scarcity of some resources and a large reduction in others, causing negative impacts throughout the ecological chain.

However, realising the future risk posed by the scarcity of these resources, guidelines and restrictive rules were created for the use of natural goods with a certain degree of control and rigour, guaranteeing the maintenance of future generations who will also need the resources to survive.

> We have reached a point in history where we must mould our actions around the world with the utmost prudence, bearing in mind their environmental consequences. Through ignorance or indifference we can cause massive and irreversible damage to the earth's environment on which our life and well-being depend. With more knowledge and thoughtful actions, we can achieve a better life for ourselves and for posterity in an environment more suited to human needs and hopes. The prospects for improving environmental quality and living conditions are broad. What we need is enthusiasm, accompanied by calmness of mind, and intense but orderly work. In order to achieve freedom in the world of nature, man must use his knowledge to create a better world in co-operation with it. It has become imperative for humanity to defend and improve the environment, both for present and future generations, a goal that must be pursued in harmony with the established and fundamental ends of peace and economic and social development throughout the world. (Stockholm Declaration on the Human Environment 1972, p.2)

The UN held a Conference on Environment and Development in Brazil in 1992, the aim of which was to assess how countries had promoted environmental protection since the Stockholm Conference in 1972. The UN Conference sparked a debate with the international community about the need for an urgent change in behaviour aimed at conserving life on Earth. The conference became known as the "Earth Summit", and one of the main results of the conference was "Agenda 21". This document represents a participatory planning instrument which openly accepts the responsibility of governments, companies and non-governmental organisations to stimulate environmental programmes and projects through policies aimed at preserving the environment. As a result, 179 countries made a commitment to contribute to preserving the environment, including Brazil.

Currently, the mass media address issues related to the environment. However, it has no intention of forming critical and active citizens, capable of giving opinions and participating in decisions involving environmental issues. The media often deal with environmental issues in a superficial way, applauding consumerism, selfishness and waste, which are detrimental to social and environmental coexistence.

> The media, which plays a fundamental role in the information age, has not given due space to the problem, placing it in isolation, in a narrative that tends to express itself as dramatic, romantic and apolitical. It also makes the mistake of placing the responsibility for the ecological crisis on the individual and behavioural level. The

> presentation of environmental disasters seems to make it difficult for viewers to establish links between the event and social dynamics (LOUREIRO, 2006, p. 25).

This brings us to the conclusion that the way in which environmental issues are dealt with is entirely linked to the socio-cultural context of a given society, i.e. they are historically constructed in a dialectical relationship between society and nature.

Critical Social and Environmental Awareness

We argue that environmental education should be developed within a set of educational practices that contribute to the development of a critical socio-environmental awareness.

> Environmental education is understood as the processes through which individuals and the community build social values, knowledge, skills, attitudes and competences aimed at conserving the environment, which is a good for the common use of the people, essential to a healthy quality of life and its sustainability." (National Environmental Education Policy - Law No. 9795/1999, Art. 1)

From this perspective, the school's main function would be to promote debate on the problems that affect the entire school community, encouraging students to reflect and produce, starting to build. Thus, no longer just reproducing what is imposed on society.

It is therefore extremely important to analyse the historical foundations of environmental education, accompanied by a reflection on the current social, political and economic context, helping to build a critical awareness.

> [...] teaching should be organised in such a way as to provide opportunities for students to use their knowledge of the environment to understand their reality and act on it, by exercising participation in different instances: in activities within the school itself and in community movements. (PCN - TEMAS TRANSVERSAIS 1998, p.190).

Another very important point is the objective and pedagogical planning, which must be well defined, avoiding isolated and temporary events. Such as sporadically organised recycling activities, which only contribute to the formation of a conservationist conscience. It is therefore clear that there are players who contribute to structuring environmental education that is not only conservationist and informative. Above all, it should form habits, attitudes and behaviours that identify, establish proposals and act to preserve the environment.

In this context, Arroyo (2000, p.70) also discusses issues related to the differences between the contents applied in teaching. He classifies environmental education in "open capacities", which permeate various areas of knowledge, such as issues of social interaction, ethics, political and moral awareness, among others, which are paramount in the educational process. In this way, Environmental Education should be seen and worked on across disciplines, establishing links of exchange between knowledge.

National Curriculum Parameter

In Brazil, the *PCNs* are guidelines drawn up by the federal government with the aim of guiding teachers through some fundamental factors relating to each subject and level of education. Their aim is to provide students with the means to use the knowledge they need to exercise citizenship. Although they are not compulsory, they serve as guidelines for teachers.

> It is hoped that the school will contribute to the constitution of a citizenship of a new quality, whose exercise brings together knowledge and information for responsible protagonism, to exercise rights that go far beyond traditional political representation: employment, quality of life, a healthy environment, equality between men and women, in short, ideals and affirmations for personal life and coexistence. (PCN - Ensino Médio, 1999, p.72).

The National Curriculum Parameter establishes that Environmental Education should be treated as an integrated, continuous and permanent educational practice, with the aim of awakening socio-environmental awareness, contributing to the formation of citizens who are aware of and active in environmental issues. With an affective action of transversal and interdisciplinary training.

Pedagogical Activities

The school must assume its role as a space for social transformation and, to do this, it is necessary for teachers to teach and learn knowledge in order to contribute to the formation of conscious citizens who are able to decide and act in socio-environmental reality. One of the clear objectives of environmental education is coherence between teaching and school practice. According to Freire apud Noal (2003), the development of an educational process is directly linked to the need to associate attitude with practice. From this perspective, it is important to involve students in interesting activities, constantly stimulating them in order to encourage their creativity and participation.

> One of the essential tasks of the school, as a centre for the systematic production of knowledge, is to critically work on the intelligibility of things and facts and their communicability. It is therefore essential that the school constantly instils the curiosity of the student rather than 'softening' or 'taming' it. It is necessary to show students that the naive use of curiosity alters their ability to find and hinders the accuracy of their findings. On the other hand, and above all, it is necessary for the student to assume the role of subject in the production of their own intelligence of the world and not just that of recipient of what is transferred to them by the teacher. FREIRE (1999, p.140).

With this in mind, we have organised some suggestions for teaching activities, the aim of which is to get secondary school students to reflect on and propose actions to intervene in the environmental problems experienced in today's society.

Activity 1: Getting to know the students.

Firstly, it's essential to find out the level of environmental awareness of each student in question. Obtaining information that will be essential when planning the calendar of activities. This can be done through the application of questionnaires on the subject, or with interactive discussions among the other students.

For the teacher-student relationship to become a teaching-learning relationship, the teacher needs to stimulate the student's creativity and participation.

Activity 2: Presentation of the theme.

At a second stage, the teacher needs to initiate the discussion, addressing the relationship between man and nature and exploring political, economic and ecological issues. Mediating the discussion in a participatory and constructive way, denying exclusively conservationist activities and informative content. To do this, the teacher needs to use educational activities that stimulate the student, such as: discussions; cine debates; lectures; conversation circles; study groups; and fieldwork.

An interesting proposal is to set up "ecological soirees" to work on environmental issues through different activities such as theatre, music, poetry, painting, etc.

Activity 3: Practical Action (Political Militancy)

After the discussions, it is the teacher's role to encourage practical actions geared towards each student's socio-environmental reality, showing the paths of political activism and sustainable development. It is also possible that in the course of the discussions, proposals for practical activities emerge, and it is up to the teacher to provide a democratic environment so that these proposals become effective actions. There also needs to be a link between the school and the community, creating channels of communication and dialogue.

Final considerations

Therefore, we understand Environmental Education as a teaching method that seeks to develop a continuous awareness in the population, in students, and consequently, in society as a whole. It should not be something expressed in a superficial and isolated way, but should become a habit present in everyone's routine, where there is a real concern to preserve the environment, through knowledge, expressed through daily attitudes, skills, motivation, in order to seek to solve existing problems and avoid future negative events.

When environmental education is aimed at a specific target audience, which in this case is high school students, it requires a certain amount of care and attention.

Any major project that ultimately aims to transform a collective view of a certain aspect or common reality also requires a lot of ambition and persistence. Naturally, when a new idea is thrown into our thinking, and into the way we articulate our daily lives, it immediately generates discomfort and reluctance; non-acceptance seems to be more obvious than change. There is even resistance from a good number of educators themselves, those who are more conservative and who don't view the issue with such admiration. The few enthusiasts are left with the task of exposing this educational model, which is based on a good relationship between the environment and man, and which will consequently benefit everyone. However, the majority of this group is limited to developing a superficial and informative concept, which ends up not having the expected effect.

Teachers therefore need to undergo appropriate training processes, with access to equivalent teaching methodologies, in order to address the issue effectively and so that the objectives set are actually achieved, resulting in a reasonable level of absorption, developing a new way of seeing the world and forming a concept of environmental citizenship.

The aim here is to view environmental education from a critical and materialistic perspective in relation to the aspects that students experience on a daily basis, within their reality and experience, seeking to bring environmental issues closer to each student. Another aspect that should be emphasised is the way in which environmental themes will be introduced, so as not to create new subjects, but to include cross-cutting themes in existing ones, which will enable integration with other subjects, avoiding disinterest and aversion on the part of the students.

Secondary education is the final stage of basic education, as defined by law. It is at this stage that the student will prove, on the basis of their knowledge, whether they are fit to enter higher education. The teachers' aim here is to ensure that the student's ability to "learn how to learn" is developed. Thus, environmental education must prioritise critical and innovative thinking, promoting social transformation and construction, forming citizens with local and planetary awareness, developing awareness, knowledge, attitudes, skills, the ability to evaluate and participate in society on environmental issues, capable of creating a critical and collective socio-environmental awareness.

Another conclusion I draw, taking into account the discussions that took place during the last National Meeting of Geographers held in João Pessoa, in the state of Paraíba (2018), is that teaching work faces several barriers that make it difficult and impossible to fully develop. Especially when the teacher decides to make a difference, rejecting traditional methodologies and uncritical content that has little or no relation to the student's reality. But even with great difficulty, teaching work is necessary, because the classroom is an important space for militancy, which is always in dispute. So I believe that the teaching professional must occupy this space, with the aim of questioning current teaching and seeking new forms of teaching and learning that are linked to the student's socio-cultural reality. This is the only way to build a critical socio-environmental awareness, capable of questioning today's society.

Now, to finalise my considerations, I'd like to bring up one last reflection, which was raised recently and which has made me think a lot about how I view environmental issues, because the work I've done follows the precepts and values of modern science, which is part of academic knowledge.

But thinking epistemologically, what are the functions of modern science? To develop technology? To improve people's quality of life? To preserve the environment? Or does it just contribute to the bourgeoisie's process of maintaining the status quo? These questions make me wonder: is environmental education developed along the lines of modern science capable of improving relations between man and nature, or is all this talk just a fallacy?

References

BRAZIL. **Law No. 9.795 of 27 April 1999.** Provides for environmental education, institutes the National Environmental Education Policy. [online] available on the Internet via <WWW.URL:http://www.soleis.adv.br/educacaoambiental.htm> Accessed on 24/04/2017.

. **Parâmetros curriculares nacionais:** terceiro e quarto ciclos: apresentação dos temas transversais. Secretariat for Basic Education. Brasília: MEC/SEF, 1998.

______ . Ministry of Education. **Environmental Education Treaty for Sustainable Societies and Global Responsibility**, [on line] available at Internet via: < www.mma.gov.br/estruturas/agenda21/ files/estocolmo.doc > Accessed on 20/04/2017.

FREIRE, Paulo. **Pedagogy of Autonomy**: Knowledge necessary for educational practice, São Paulo: Paz e Terra, 1999.

GIL, Antônio Carlos. **How to prepare research projects**. 4. ed. São Paulo: Atlas, 2002.
LOUREIRO, Carlos Frederico Bernardo. **Social theory and the environmental question:** presuppositions for a critical praxis in environmental education. In: LOUREIRO, Carlos Frederico Bernardo; LAYRARGUES, Philippe. Pomier; CASTRO, Ronaldo Souza de. (Org.). Sociedade e meio ambiente: a educação ambiental em debate. 4. ed. São Paulo: Cortez, 2006.

SILVA, Ângela dos Santos Maia Nogueira da. **A Look at Environmental Education in High School:** Practising Theory, Reflecting Practice. Florianópolis: UFSC, 2003.

2. LIBERTARIAN GEOGRAPHY:
THE QUEST FOR ANTI-AUTHORITARIAN EDUCATION

Introduction

During my teacher training process, I was able to learn about and identify the various approaches and perspectives that have influenced and still influence geography teaching in public schools. Through the discussions developed in the classroom and, above all, through the direct contact with the school provided by the supervised internship courses and PIBID.

So I realised that the geography teachers who teach in the public schools I had the opportunity to attend are still reproducing traditional and authoritarian teaching practices and methodologies.

From this perspective, I decided to develop this work with a focus on the libertarian perspective, which seeks to develop truly free and liberating teaching. Even though it is impossible to develop an anarchist pedagogical practice within the current system in which we live, the development of a methodology guided by the libertarian pedagogical trend is important because it goes against the traditional education in force and creates new possibilities for teaching and education.

Libertarian pedagogy, like many other ideas that criticised the prevailing order and ideals of the time, derives from the French Revolution (1789). But throughout history, libertarian pedagogy has taken on different forms, with different ideological political values. Anti-authoritarian pedagogy is the focus of this discussion and brings together the work of some anarchists and social activists, who work in favour of an education that seeks autonomy, freedom and the reflexive development of the subject.

It is important to emphasise that libertarian pedagogy conflicts with the current "liberal" thinking that dominates Brazil's public schools. This is based on traditional liberal values - such as obedience and uncritical training - geared towards the labour market. Hence the contradictions between "liberal" education and "libertarian" education, which seeks to end the asymmetries of power that make the teaching and learning process an instrument of domination.

General Objective

Introducing libertarian thinking into geography teaching in public schools

Specific objectives

- Identify the shortcomings of current geography teaching
- Discuss the theoretical bases that underpin the discipline
- Point out the need for innovative teaching
- Present the main characteristics of libertarian pedagogy

- Encouraging practices of pedagogical transgression

Working Methodology

This work is based on bibliographical research aimed at achieving the proposed objectives. Initially, a bibliographical review was carried out with the aim of gathering works by some authors who discuss the teaching of geography within the framework of libertarian education.

The literature review is qualitative in nature and was carried out through systematic reading, with summaries, abstracts and syntheses, seeking to highlight the pertinent points of each author studied.

Bibliographical research is important, according to Marconi and Lakatos (2003), because it involves all the bibliography that has already been made public in relation to the object of study, including books, magazines, scientific articles, texts in general, among others, putting the researcher in direct contact with what has already been discussed and published on the subject.

3. Results and Discussion

3.1 Current geography teaching

Today in Brazil, the field of education research already concentrates a great deal of dense material. Presenting work from the most diverse areas of knowledge linked to teaching, such as psychology and sociology of education, which uses other knowledge to understand educational processes.

However, even with the advances in scientific production in the field of education, it is possible to identify in practice that traditional teaching methodologies are still used in almost hegemonic fashion in the country's public schools. Especially in relation to the teaching of geography, which still follows the positivist model introduced in Brazil in a period prior to the proclamation of the republic. But to understand the current teaching of geography, it is necessary to revisit its trajectory throughout the country's school history. Analysing the theoretical and methodological development of geography teaching within teaching practice in different contexts.

In Brazil, geography as scientific knowledge first emerged in the context of schools. Geography was included in the influential and important basic education teaching centre of the time, the "Colégio Pedro II" (1837), and the subject became compulsory in the country's other schools. From its introduction into schools, geography began to have a descriptive function, based on cartographic representations that sought to valorise the territory and create a relationship of respect and love for the image of the homeland. Within this context, Geography became official in schools with the aim of training future patriots willing to die for their country if necessary.

Throughout its development as a school subject, geography was reduced to encyclopaedic, mnemonic teaching, with lists of names to be copied and memorised. Even during this period, as there was no higher education in geography, textbooks became a reference in Brazilian geography and dominated the educational field. But it wasn't until the 20th century that various works began to appear that challenged the way geography was produced and taught. Thus emerged libertarian conceptions based on anarchism, critical conceptions based on Marxism, and later other approaches with different perspectives. Over the course of time, they were appropriated by education with more or less success, depending on the social, economic and cultural context of each society. In Brazil, these concepts were more widely accepted in private schools, which catered for the dominant classes. While in public schools, traditional conceptions of teaching and learning continue to be used.

According to Libâneo (1989), traditional pedagogy in Brazil is geared towards the children of workers, directing teaching towards supplying labour for the job market, efficiently transmitting precise, objective and quick information. Therefore, the discussion on the theoretical-methodological issue of teaching geography is, consequently, a debate on the theoretical foundations that underpin the discipline.

In this way, it is possible to see that geography in public schools is incapable of producing the ethical, rational and political effects committed to society because this institution is antagonistic to these purposes. Libertarian pedagogy, on the other hand, takes a different direction, proposing critical, anti-authoritarian teaching that seeks to provide the necessary means for building autonomy and self-management.

3.2 Libertarian pedagogy

But in order to understand libertarian ideas, which have always been present since antiquity, it is necessary to refer back to the history of education. In the vast Greek literature we can find elements of utopian anarchism in Plato, considered the first Greek educator. From this perspective, the development of teaching practices and the attempt to develop a system of education have, throughout history, taken divergent political and philosophical paths, which have materialised in education developed under bourgeois liberal values and others based on libertarian principles that question the power relations in force in society, such as libertarian and anti-authoritarian pedagogy.

Sobreira (2009) characterises anti-authoritarian pedagogies:

> Anti-authoritarian pedagogies can be called socialist because they defend the principles of radical democracy, self-management, the autonomy of the subject committed to collective results, with no contradictions between the objectives of education, as is the case with liberal education, which selects and separates by gender, class and other criteria of inequality. (SOBREIRA, 2009, p.40).

Understanding the importance of searching for the essence of each teaching practice, Suissa (2006) exposes the essential value of autonomy, self-management and freedom that libertarian pedagogy carries. Denying the meritocratic values, based on vocational utilitarianism, defended by liberal pedagogy.

> In the field of education, libertarian pedagogy is defended by many as an alternative for quality emancipatory practice, based on principles of self-management, affirmation of freedom and federalist principles of government. These characteristics, when applied to teaching, would cause a rupture in the liberal paradigms pervaded by the dualistic teaching structure. (CHERUBINI, 2010, p.8)[.]

Libertarian theory assumes that no one has the right to impose what another person should learn. Anarchist theorists therefore advocate self-regulated learning, the non-fragmentation of knowledge and respect for each person's pace. In addition, all decisions involving the process of building knowledge are decided in assemblies. From this perspective, anarchists argue that education is part of the path to transforming society, but question whether the best environment for this process to take place is actually the school. Most argue that not even an institutionalised school is capable of producing an anarchist education, because its function is antagonistic to libertarian purposes. But within a school with a libertarian bias, it is possible to put into practice some anarchist precepts of education, such as anti-authoritarian and self-regulating teaching, anti-sexist education, the autonomy of the subject, learning in freedom and values.

Given this statement, the geography that is currently being taught in Brazil has no place within anarchist education because it is fragmented and lacks articulation with other areas of knowledge. In anarchist literature there are three perspectives on education: education along the lines of Stirnean thought, education within the school in Bakunin and the thesis of anti-pedagogy, which advocates de-schooling.

Bakunin and other anarchists have developed the thesis that within the institutionalised school environment there can be an anarchist education, but other libertarian theorists and activists don't accept the idea and defend the creation of knowledge-building spaces independent of the state as the only way to achieve a true libertarian education.

Within this perspective are the advocates of the de-schooling of society as a whole. According to Sobreira (2009), de-schooling theorists criticise the school entity and advocate a new way of thinking about school, without walls, without bars, denying the State and the other agents who dominate and dictate from the top down how the teaching process should be.

> The school system also rests on a second great illusion, that most of what is learnt is the result of teaching. Teaching is true, it can contribute to certain kinds of learning under certain circumstances. But most people acquire most of their knowledge outside of

school; in school only while it has become, in some rich countries, a place of confinement for an ever longer period of their lives. (ILLICH, 1973, p.36).

From this perspective, de-schoolisation aims to destroy all formal schools and create in their place another model of education that is not cut off from society as is the current school-centred educational model. Such action first depends on profound changes in society and there is no recipe for how this should happen. There are only a few experiences that can serve as a basis for this change.

In Brazil, for example, the labour movement was very important for the dissemination and consolidation of some anarchist precepts. Within the movement, anarchism was present in the last decade of the 19th century and the first three decades of the 20th century. So much so that in 1906 the first Brazilian Workers' Congress (COB) took place, with a strong presence of anarchists. This congress was important because it resulted in the construction of educational actions committed to the schooling of workers in Brazil. The schools under the influence of the unions took advantage of the libertarian education proposals drawn up in Europe, mainly in Spain, adding up to more than 40 schools and study centres. An example of this proposal is the Young Idealists Women's Centre, which focuses on feminist thinking in Brazil, another real experience that is little researched within universities. Like the People's University, founded in 1904, which had a bold plan and advocated anti-state education aimed at working people. However, the anarchist influence was strongly attacked in the following decades, mainly by the media of the time, the newspaper and the radio, which exerted a strong influence.

The official press of the time was opposed to the rapidly growing anarcho-syndicalist movement. Several newspapers linked to the Church attacked the anarchists and criticised the schools they had founded. In addition, the State, through government interventions, led to the closure of several schools, such as the Modern Schools in São Paulo and São Caetano, which followed the parameters established by a group of anarchist activists, pedagogues who exerted a strong influence in Spain, among them Francesc Ferrer and Anselmo Lorenzo. According to Ferrer's model, the aim of the school was to educate the working class in a rational, secular and non-coercive environment.

3.3 Libertarian pedagogy and geography

The rapprochement between libertarian education and geography is made through the writings and reflections of anarchist geographers Elisée Reclus and Piotr Kropotkin. Two important geographers who, because of their ideological convictions, were often ignored by modern science and the state.

In her reflections on teaching, Elisée Reclus demonstrates her ideas both in relation to geography and her commitment to anarchism, affirming a conception of education that seeks maximum freedom for the student and breaks with the teacher's relations of domination and authoritarianism.

In 1885, Kropotkin wrote and published the text "What geography should be", which proposed the reorganisation of geography and its contact with other sciences. He was a pioneer of the proposals for multidisciplinarity through a reorganisation of scientific branches, always in search of more spaces of freedom for thought.

So even with the countless works carried out by anarchist educators and researchers, the teaching of geography from a libertarian perspective has gained little acceptance, given the positivist current, based on logic and memorisation methods, and the critical current, based on dialectical historical materialism. This lack of expression may be due in part to the great difficulty of including anarchist ideas within an academic discipline created and organised according to pre-established criteria by hierarchical educational institutions.

In this respect, Sobreira (2009) contextualises it well, questioning the importance that is still attributed to the teaching of cartography. As if the students had no spatial notion and cartography arrived as a saviour, as a form of cartographic literacy. But in reality, if students don't understand spatial location issues, it's due to a lack of freedom in schooling and not because the students are undisciplined or stupid. As Suissa (2006) rightly writes, when he exalts the value of education for freedom, defending critical dialogue and encouraging creativity.

From this perspective, the fundamental principle of anti-authoritarian education revolves around the assumption that no one has the right to define what another should learn, but rather to encourage self-determination in learning committed to the community and to oneself, which is why there is no point in isolated and fragmented curricula or subjects, as is currently the case within public education institutions.

3.4 Pedagogical transgression

Pedagogical transgression seeks, through teaching practices and methodologies, to break with some level of tyranny present in the school. Such as the teacher-student relationship, the organisation of the curriculum and content, as well as school management.

However, these attitudes and practices are not capable of changing the reality of schools, as they only represent the will of some teachers not to reproduce the traditional teaching in which they are inserted. But only a libertarian school project, supported by society, is capable of producing innovative and liberating teaching. Giving teachers the freedom to work freely without institutional demands, and above all, encouraging the adoption of libertarian practices and methodology.

But given the current reality in schools, what should teachers do? Continue reproducing traditional teaching? I don't think so, but we should innovate. We just can't fool ourselves into thinking that we're going to change society with innovative methodologies alone. Political activism is necessary inside and especially outside the classroom, because with the current educational structure we have, the teacher inside the classroom is not capable of saving the school. It's important for the teachers' union to take a stand not just for salaries, but for another educational structure, without walls, without rules, without prescriptions where the teacher and the student are really taken into account.

As a result of current traditional teaching, which reproduces inequalities, the use of libertarian pedagogy is necessary, especially within geography, which is a science marked by logic and spatial domination. It seeks to study and explain the natural and social relationships that produce and organise space at all time scales. But why should we teach and learn geography? Is the science of geography really that important? And what should we use it for?

Libertarian pedagogy does not advocate teaching any subject the way it is done, imposed, fragmented and without articulation with other areas of knowledge. For students to learn geography, they have to want to learn, they have to be attracted to knowledge. And that's the role of the teacher, to create a favourable environment for the student to seek knowledge themselves, without a relationship of authority.

The teacher in this case would be a mediator or facilitator of the student's knowledge process. Creating forms of interaction with the environment that provide meaningful learning. To do this, it is important that the teacher listens to the student and does not just use directive, pre-established content. Using the principle of non-directivity, which is based on the individual's self-learning, helps to build autonomy in learning. Non-directivity is often confused with studying only what you feel like studying. But this is a mistake, because directness is also important within the teaching and learning process, the fundamental problem lies in the fact that what is taught is imposed. It doesn't allow the teacher to create new ways of teaching and learning, nor is it interested in the socio-cultural reality of the students.

That's why it's important for the teacher to get in touch with the student's reality and, through this contact, facilitate the process of building knowledge. Providing an environment that stimulates learning, in this logic the teacher is not a transmitter of knowledge, but a facilitator who joins the group of students.

In the classroom, transgression can simply mean not following a specific textbook or pre-established content. But for this to happen, it is also important for the teacher to be familiar with all the geography content and to have a wide range of bibliographical material available, working on the same topic in different ways.

Some teaching approaches and practices are easier to carry out, but are nonetheless important because they create new learning possibilities, such as fieldwork, research projects, science fairs, themed soirees and support for the formation of student guilds.

Of course, if the teacher decides to use a libertarian approach, he or she will face a number of challenges, such as the number of students and the limited time available. These end up pushing the teacher towards traditional methodologies, which are already absorbed by the students and don't require as much time and planning. What's more, pedagogical transgression can lead to teachers becoming unemployed, which is why many choose to continue reproducing current teaching, since in the end it's the teacher who loses out.

4. Final considerations

In my research, I was able to make some considerations in relation to the teaching of geography and especially in relation to the libertarian perspective of society and education. Which in Brazil currently has little expression both within militant movements and within schools.

The neglect of the libertarian current of thought within geography may be a consequence of the fact that there was no successful national school geography in the country with the aim of confronting the interests of the state and the bourgeois elite. Even with the presence of anarchist ideas within the Brazilian syndicalist movement at the beginning of the last century, Brazilian geography remained rigid, following positivist ideas, both in the field of research and in school education.

With all this, it is possible to conclude that libertarian ideals here in Brazil had a period of ascension, but were soon strongly attacked by the Church and especially by the State. Today there are still a few educational institutions with a libertarian bias, but with little support and hardly any expression. This ends up hindering the concrete development of a libertarian, anti-authoritarian education that is committed to popular education.

But even with this impossibility, the use of libertarian methodologies and positions is necessary, as it conflicts with the current school institution and creates new teaching possibilities. It is also an important mechanism for social criticism, proposing another way of looking at society and the world. This is why the discussion and use of libertarian practices is important, because it is only through the propagation of these ideals that libertarian education can one day truly exist in Brazil.

References
CARRÃO, Paulo. Vitor. Miranda. **Anarchism and Education.** 1992. Dissertation (Master's in Education) 200f. Fluminense Federal University, Niterói, 1992.

CHERUBINI, Iris Cristina Barbosa. **Libertarian pedagogy:** a historical look at the limits and possibilities of its implementation in Brazilian public schools. State University of Western Paraná. Cascavel, Paraná, 2010.

CORDEIRO, Gisele do Rocio; MOLINA, Nilcemara Leal; DIAS, Vanda Fattori (org). **Guidelines and**

practical tips for academic work. 2 ed. - Curitiba: Inter Saber, 2014.

GALLO, Silvio. **Anarchist Education**. Piracicaba: Artmed, 1995a.

GIROUX, Henry. **Critical theory and resistance in education**: beyond theories of reproduction. Petrópolis: Vozes, 1986.

ILLICH, Ivan. et. al. **Education and Freedom**. São Paulo: Imaginário, 1990.

KAERCHER, Nestor: Geography is our everyday life. In: **Geografia em sala de aula**: práticas e reflexões. Porto Alegre: Editora Universidade Federaldo Rio Grande do Sul and Associação dos Geógrafos do Brasil, 1998.

KASSICK, Neiva. Beron; KASSICK, Clovis. Nicanor. **A pedagogia libertaria na história da educação brasileira**. 2. ed. Rio de Janeiro: Achiamé, 2004.

MELO, Adriany de Ávila; VLACH, Vânia R. F. **Uma Introdução à História da Geografia Escolar Brasileira**. In: Proceedings of the VIII Meeting of Latin American Geographers, Santiago, Chile, 2001

SOBREIRA, Antônio Elísio Garcia. **Anarchist pedagogy and geography teaching**: conquering freedom quotas. 2009. 358 f. (Thesis in Geography) - UNESP Presidente Prudente, São Paulo, 2009.

3. URBAN PROCESSES IN THE PRODUCTION OF SPACE - AN ANALYSIS OF THE TUPÃ NEIGHBOURHOOD IN ITUIUTABA-MG

Introduction

Nowadays, the production of urban space has been characterised by relationships, processes and actions, making it difficult to establish what this space is. Thus, it was based on Marx's idea of the annihilation of space by time that Lefebvre raised the possibility of theorising social space under the assumption of its production.

> Space is not only part of the forces and means of production, it is also a product of these same relations. Lefebvre observes that, as well as there being a space of consumption or, in other words, a space as an area of impact for collective consumption, there is also the consumption of space, or space itself as an object of consumption (GOTTDIENER, 1993, p. 129).

According to Smith (1986), we start from the principle that today economic expansion does not develop through absolute geographical expansion, but through the internal differentiation of geographical space, understanding the production of space as an unequal process of capital reproduction. From this perspective, this study is based on an analysis of the real estate dynamics around the UFU campus in the city of Ituiutaba, with the aim of identifying and analysing the processes of reorganisation and restructuring of urban space. As a result, we present a map of the current spatial organisation of the neighbourhood in which the campus is located, highlighting the transformations that have taken place as a result of the new development.

Effects of building the campus

The Ituiutaba City Council, in partnership with the private sector, donated a 500,000m plot of land^2 near the Tupã neighbourhood for the construction of the new UFU campus. The campus was inaugurated in 2012, starting its activities in the new unit. The installation of the campus and the consequent increase in the flow of people triggered a process of urban restructuring, altering the local spatial dynamics.

> The Tupã neighbourhood, prior to the rumours of the university's establishment, was seen as a typical peripheral neighbourhood, lacking essential services for the population, such as quality public transport, sewage systems, asphalt and public services in general, in other words, just another neighbourhood characterised by a lack of planning (SILVA and LOBODA, 2013, p. 111).

It is worth highlighting the agents that have contributed to the process of urban restructuring, such as real estate agents who, through the internal differentiation of urban space, determine the location that will be valued.

According to Silva Júnior *(apud* RESENDE, 2011), "to speculate is to hold back, to keep something in the probability of making an advantageous bargain as soon as the need for profit arises, when the value to be received from the bargain would be much higher than the price of the asset". In the case of the Tupã neighbourhood, according to Silva and Loboda (2013), since the announcement of the location of the new campus there has been a process of real estate speculation, reflected in the appreciation of land, increasing rents and opening up space for new buildings. The process of gentrification is commonly linked to property development:

> [...] access to adequate housing conditions is linked to income and, as urban space is appropriated for speculation, the low-income population is rewarded by the alternatives presented to them: seeking ways to occupy part of the territory in inadequate conditions or committing to long loans and living further and further away from the valued areas (COMPANHIA DE HABITAÇÃO DE LONDRINA, 2011, p. 51).

In the case of the Tupã neighbourhood, the gentrification process benefited students to the detriment of the area's former residents. Property speculation was so successful that there was no need to invest heavily in infrastructure in the neighbourhood: asphalting was enough for not only flats but also buildings of considerable height to be erected to meet the demand for university housing. There are no squares, Goiabal Park is in a precarious state, and the neighbourhood has no leisure options for residents, including university students. Even so, as we'll see in the mapping of the neighbourhood, several homes have been abandoned and others have been turned into student housing, which exposes the expulsion of the neighbourhood's former residents.

Methods and materials

Firstly, a bibliographical survey was carried out on the subject, consulting various sources. Secondly, a cartographic survey was carried out to find out the boundaries and roads of the Tupã neighbourhood.

After identifying the area of the neighbourhood, fieldwork was carried out "in loco", observing, analysing, photographing and noting relevant aspects in each block of the neighbourhood, from vacant lots, buildings, asphalting, among other structural aspects. Finally, the information collected was analysed, highlighting changes in property dynamics compared to the work carried out previously by Loboda and Silva, which coincides with the year UFU was set up.

The methodology for the work was developed in such a way as to make it possible to map the entire neighbourhood. Initially, a sketch was made of the neighbourhood's street layout, so that the location of each of the pre-established types of plots in the neighbourhood could be drawn by hand. These plots were classified as

1) **Residence**, which includes the simple houses in the neighbourhood;

2) **Plots with more than one residence** include flats and houses built sequentially on the same plot to the same standard, often built to meet the demand for student housing at UFU (LOBODA and SILVA, 2012), and plots with separate houses at the front and back of the plot;

3) **Vacant plots: plots that** appear to be untouched and identifiable. Plots that were not fenced or whose boundaries could not be identified through observation were counted as just one plot;

4) **Built-up plots are** plots with abandoned or ruined buildings, allowing the perception that the plot, although no longer in use, once had a function in some previous time frame;

5) **Lote Plantado**, plots commonly used by the population for planting various crops. Through information gathered from local residents, it was possible to find out that these plots often don't belong to the people who grow the crops there, and the planting is done without the plot owner's authorisation;

6) **Chácara**, plots with visible planting, whose use is based on urban agriculture; although it was not possible to identify whether the owners aimed to make a profit from the planting practice, we made it clear that the denomination is only intended to associate the use of the plot with the practice of planting (mainly) and animal husbandry in the neighbourhood studied, with no direct link to leisure or the sale of products from the aforementioned activities;

7) **Under construction,** any plot with construction worker activity or with a newly built structure;

8) **Building**, any structure with 2 or more floors and 4 residences;

9) **Commerce**, markets, bars, bakeries, all kinds of private services offered in the neighbourhood through a physical shop.

Once the information had been gathered, we moved on to digitising the map. Using QGis software, it was possible to reproduce the structure of the neighbourhood, and each street/plot was transformed into a line/polygon. The OpenLayers plugin allowed a satellite image to be used in the background, making it easier to delimit plots in the neighbourhood and also to structure the blocks. Despite being very close to the real thing, the boundaries cannot be used as a reference for the size of the plots, since the idea of drawing the plots according to the satellite image is only to better represent the distribution of each type of plot along the blocks. Furthermore, this representation was not based on data on plot sizes in official documents, but was the result of the group's observation and deduction in relation to the image used as a base, taken from Google Earth.

After producing the digitised sketch, we began to analyse and compare it with the material previously produced by Silva and Loboda in 2012, in order to observe the changes that have occurred since then in the configuration of the neighbourhood, allowing us to identify the real estate dynamics and urban processes underway in the study area.

Results and discussion

Thus, we introduced the basis of our research: the characterisation and mapping of the urban structure of the Tupã neighbourhood (Figure 2) and comparison with the results obtained in 2012 (Figure 1), surveying and characterising fundamental elements for an analysis of the process of producing urban space around UFU.

Figure 1: Summary map of the Tupã neighbourhood (2012)

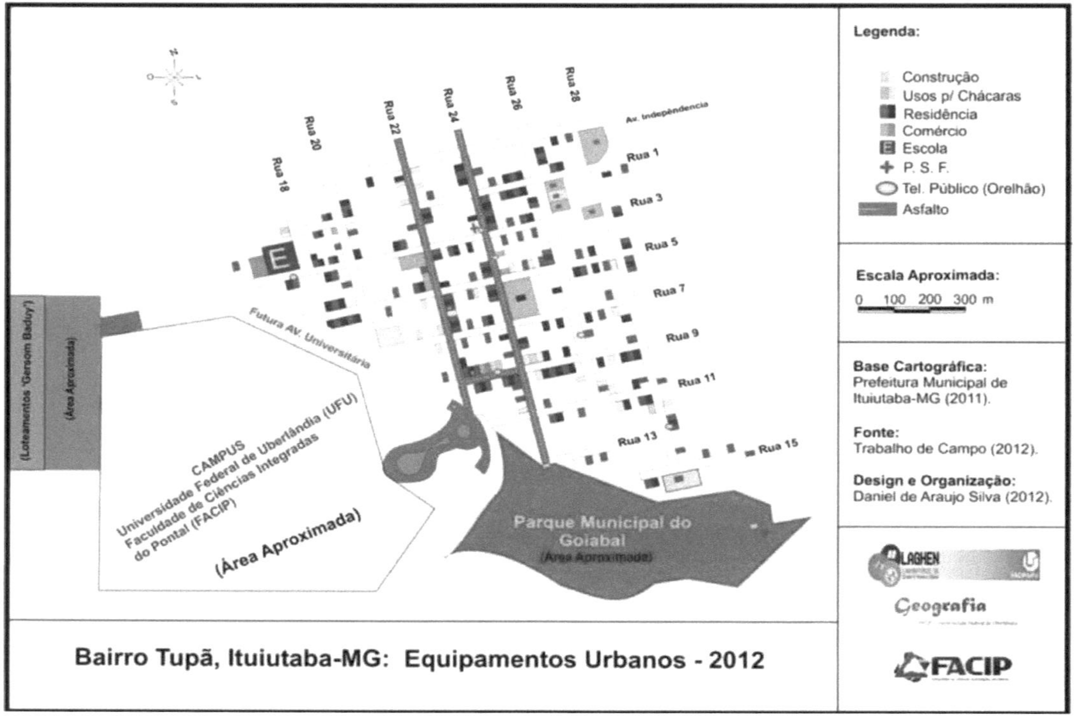

Org.: SILVA, Daniel de Araújo (2012).

Figure 2: Summary of the Tupã neighbourhood (2017).

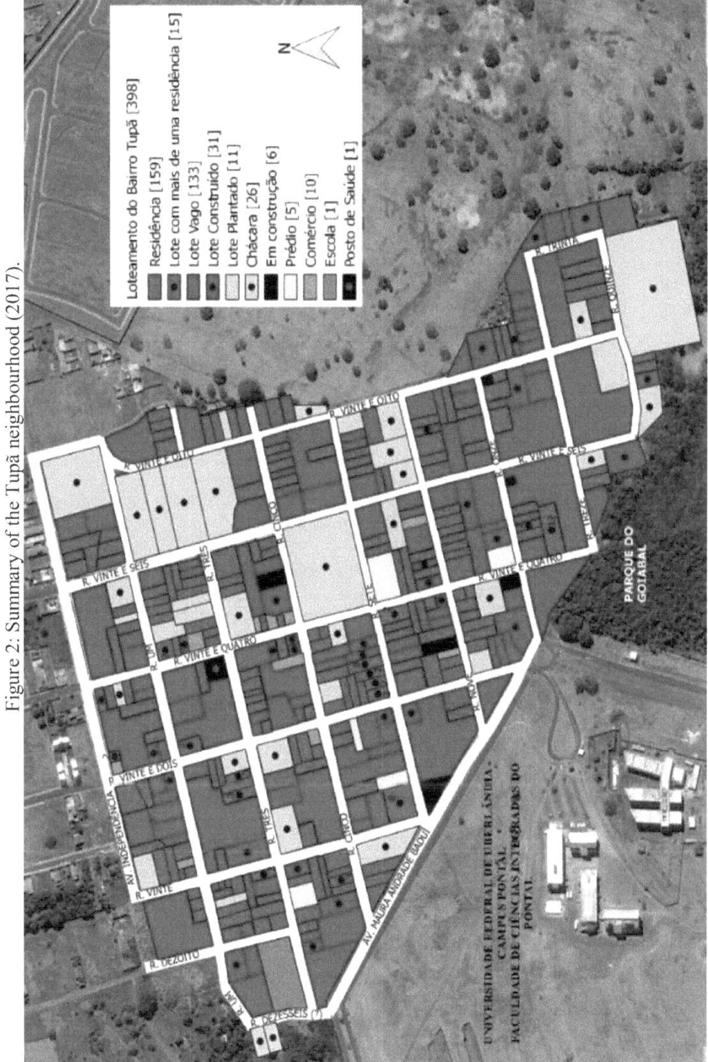

Elaboration: LEITE, David (2017); Organisation: LEITE, D.; CAVALMORETTI, R.; PENARIOL, R. (2017)

Trade, Goods and Services

Commerce in the neighbourhood is still geared towards meeting immediate local demand. The private services on offer are those that meet quick needs, with limited stock and at a price above that of the market. This is due to the fact that most local traders replenish their stocks at the Mart Minas atacarejo and other local supermarkets, which charge market prices. In addition, immediate demand also affects the price: the distance from larger shops that offer a certain product and the urgency of obtaining this product result in higher prices for those who offer it. We can highlight, when comparing the previous study carried out by Silva and Loboda, the opening of a thrift store, which is the first to offer clothing and accessories, which did not previously exist in Tupã, as well as a mechanic's workshop and a blacksmith's shop.

Trade in the neighbourhood has increased in terms of the variety of sectors since 2012, but has been facing problems recently. We can cite the economic crisis that hit the country, which began in 2014 and was widely reported in the mainstream media, as a possible reason for the difficulty some businesses are having in operating. However, demand for certain services is low in the neighbourhood, and the few local businesses, according to information from the traders themselves, are struggling to stay in business. This is due to the fact that a portion of immediate needs, especially food, are met by the strong presence of delivery services, which serve the neighbourhood with a certain agility, given that a portion of this market is located in nearby neighbourhoods, such as Alcides Junqueira, which offers delivery services, for example, of ready-to-prepare meats (such as Casa de Carnes Maravilha, often used by local residents), and the Nova Ituiutaba housing estates, which seem to have found an alternative source of income in the delivery system, given that the city as a whole doesn't have enough jobs to match demand. We are committed to a future survey to confirm this information, since what has been said here is the result of observations made by the group, as well as brief interviews, routine conversations and daily contact with the reality of the neighbourhood and local residents.

As for public services, they remain the same as in the previous survey: a health centre to meet local demand for medical appointments and referrals, as well as first aid, and the Rosa Tahan Municipal School, whose classes range from the 1st to the 5th year of primary school.

Constructions, Buildings and Homes

One of the most visible characteristics of urban space is the significant presence of different spatial forms, materialised in different land uses. Thus, according to Corrêa (1989), it is not difficult to witness the opening of an urban road, the construction of a building or the demolition of an old house. In short, the city's space is being produced and organised all the time.

In this context, we highlight the inauguration of new buildings in the neighbourhood, which are clearly aimed at meeting the demand for housing from UFU students. Even some city dwellers are moving closer to the university, given the poor urban mobility and precarious public transport in the area. One example is the San Petrus Building (figure 3), near Goiabal Park. According to the owner, only one flat was sold off-plan; all the others were sold before the property was even built. Today, according to the owner, practically all the flats are rented out to UFU students.

Figure 3: San Petrus Building.

Source and Organisation: CAVALMORETTI, Personal Archive (2017).

The construction of the building led to the demolition of the house that was situated on the plot at the back of where the structure was built, joining several other abandoned houses in the neighbourhood (figure 4). Some houses are still in good condition, even after being abandoned. While some have perished, others are in reasonable condition, at least as far as we could see.

The landscape resulting from these events reinforces the assertion that the process of gentrification in the Tupã neighbourhood is underway, and is the result of the presence of UFU, both as a physical structure and as a promoter of changes in the local real estate dynamic, which was previously stagnant in the form of a peripheral neighbourhood whose land was devalued and geared towards fixed residences, and is now in the form of a provider of temporary housing for students, since the turnover in renting properties for this purpose is high.

Mobility and Accessibility

The street and its extensions must reinforce its character as a place of relationships, which guarantee not only the vitality of the environment, but also its sustainability and maintenance. The concept of mobility is related to the movement of people in the urban space, which should facilitate their journey, with clean, safe, tree-lined streets, wide pavements, adequate lighting, signposting and full accessibility. The Tupã neighbourhood has poor urban mobility, as the streets have no signs or adequate lighting, there are almost no pavements, and those that are present are almost always irregular. The main access streets to UFU (figure 5) are the best structured, but still have irregularities.

Figure 4: One of the main access streets to UFU-Pontal in the Tupã neighbourhood.

Source and Organisation: CAVALMORETTI, Personal Archive (2017).

With the installation of the UFU-Pontal campus, the neighbourhood was increasingly paved, prioritising the streets that lead to and around the university. Almost all the streets in the neighbourhood are paved, but only a few streets on the edges of the neighbourhood are still dirt, such as Street 15 (figure 6).

Figure 5: Street 15 in the Tupã neighbourhood.

Source and Organisation: CAVALMORETTI, Personal Archive (2017).

When it comes to urban accessibility, it's essential to respect the physical and sensory diversity between people and the changes our bodies go through, from childhood to old age. We should always think about inclusion, with ramps, wider pavements, signs on pavements for the visually impaired and cycle paths. It is clear that the neighbourhood in question does not provide any access for people with walking difficulties, as can be seen in figure 7.

Figure 6: Access ramp for the disabled in the Tupã neighbourhood.

Source and Organisation: CAVALMORETTI, Personal Archive (2017).

Gentrification in the Tupã neighbourhood

The process of gentrification is linked to the organisation of urban space according to the needs of the mode of production and the social class that dominates the economy of a society in a given historical period. According to Smith (1986), the production of geographical space is an absolutely unequal process and gentrification is part of the internal differentiation of geographical space in the urban school.

In this way, we understand the process of gentrification not just as a simple case of replacing low-cost housing with housing for the wealthy, but as an element in the process of reorganising the space of production, circulation and consumption of goods.

We start from an understanding of the processes of urban restructuring and reorganisation that affect a region or neighbourhood by altering the dynamics of the local composition, such as new commercial premises or the construction of new buildings, adding value to the region and affecting the local low-income population. Thus, by analysing the evolution of the neighbourhood over the last five years (on average), we can see a process of urban restructuring produced by real estate agents as a result of the installation of the UFU campus, which has caused a change in the urban structure, increasing the value of the land, excluding the poor population and paving the way for a new real estate dynamic, aimed at meeting student demand. The installation of the campus has brought few infrastructural improvements in relation to the impacts it has caused on the routine of the local population, since the government has not even implemented a policy to assist the families who live there. But on the other hand, it is strongly involved in the production of this space, intervening directly in the construction of new low-income housing estates in the outlying areas, changing the whole dynamic of the region.

Final considerations

In our studies of real estate dynamics and urban processes in the Tupã neighbourhood, we can see that urban space is dynamic and its forms are being built, destroyed and rebuilt all the time. As urban space is fragmented and articulated, changes in a particular fragment influence the whole of urban space.

In relation to the installation of the UFU campus, in addition to the transformations of pre-existing forms and the daily dynamics of the neighbourhood itself, the areas surrounding the campus have undergone a continuous increase in value, which enhances their appropriation/transformation by real estate capital. Thus, the actions of real estate capital in recent years in the vicinity of UFU are an indicator of the consolidation of this neighbourhood as a place designed to meet student demand.

Understanding the historical evolution of the neighbourhood helped us to understand its current dynamics by analysing its occupation, its structuring and its urban growth linked to the installation of an important higher education institution.

References:
PONTAL CAMPUS. Available at <www.facip.ufu.br>. Accessed on: 7 July 2017.

IBGE - **Brazilian Institute of Geography and Statistics**. Cities. Ituiutaba. Available at <http://www.ibge.gov.br/cidadesat/xtras/perfil.php?codmun=313420>. Accessed on: 7 July 2017.

OLIVEIRA, Bianca Simoneli de. **Ituiutaba in the tijucana urban network**: socio-spatial (re)configurations from 1950 to 2003. 208f. Dissertation (Master's in Geography). Institute of Geography, Federal University of Uberlândia, Uberlândia, 2003.

SILVA, Daniel Araújo; LOBODA, Carlos Roberto. **The process of producing urban space:** Mapping and characterising the Tupã neighbourhood in Ituiutaba-MG, Special issue, Barra do Garças-MT, Revista Eletrônica Geoaraguaia, p. 108 - 127, 2013.

RESENDE, Ubiratan Pereira de. **Quality of life, urban environment and property speculation:** a study on the implementation of Cascavel Park, in the southern region of Goiânia. Proceedings of the II SEAT - Symposium on Environmental Education and Transdisciplinarity UFG / IESA / NUPEAT. Goiânia, May 2011.

LONDRINA (PR). **Law No. 1.008, of 26 August 1965**. Provides for the creation of the Londrina Housing Company. FL, Londrina, 11 September 1965. Available at: <http://goo.gl/R1V3gG>. Accessed on: 18 July 2017.

SMITH, Neil. **Uneven development**: nature, capital and the production of space. Rio de Janeiro: Bertrand Brasil, 1988.

RUBINO, Silvana. Urban Ennoblement. *In:* **Plural de Cidade: novos léxicos urbanos**. FORTUNA, C.; LEITE; R. P. (org.). Coimbra, Ed. Almedina, 2009, 344 pages.

4. THE IMPORTANCE OF THE INSTITUTIONAL TEACHING INITIATION SCHOLARSHIP PROGRAMME - PIBID: Supporting educational practices

1. Introduction

As a result of scientific advances and changes in living standards, the world is constantly shortening distances. This shrinkage shows us that through innovations in transport and communications, time is annihilating space. In this way, space can be appropriated by science. Modernism alters the meaning of space and time: they should be organised and dominated to facilitate and celebrate man's liberation (HARVEY, 1993). In the light of this reflection, it can be seen that the teaching-learning relationship, which is human but complex, is part of the construction, behavioural and professional fulfilment of an individual. Thus, the analysis of teacher/student relationships encompasses intentions and interests, with this conviviality being the object of consequences, since education is one of the most important sources of behavioural development and an element that adds values to members of the human species.

According to Gomes (2006, p.233), a pedagogical practice needs to have its own dynamics, enabling the exercise of reflective thought, leading to a political vision of citizenship and capable of integrating art, culture, values and interaction, thus fostering the recovery of subjects' autonomy and their occupation of the world in a meaningful way.

According to Gadotti (1999), in order to put dialogue into practice, educators must not place themselves in the position of being the holder of all knowledge and information. Rather, they must place themselves in the position of not knowing everything and always being willing to receive the knowledge of each student according to their experiences, recognising that even an illiterate person is the bearer of empirical knowledge through lived experiences.

According to Brait, et-al (2010), when conducted in this way, the practice of learning becomes more interesting the moment the student feels included and contemplated by the attitudes and methods of motivation in the classroom. Since the pleasure and involvement of learning is not a feeling that arises spontaneously in students, it is not a task that they fulfil with satisfaction most of the time because it is seen in some cases as an obligation.

In order for students to be successful, they need to be awakened to curiosity about the world in which they live, seeing and understanding the processes that make it up and of which it is a part, the functions, consequences and, consequently, how to interconnect them, so that there is greater dialogue between the existing reality and the world with the student in training, accompanied by the teacher.

According to MEC (2014), the Institutional Teaching Initiation Scholarship Programme (PIBID) was created with the aim of valuing teachers and supporting undergraduate students at public (federal, state and municipal) and community higher education institutions. The functions and benefits of this programme therefore go beyond the tangible and the

This is based on the evolution of the individual as a teacher and consequently reflects on the formation of an individual citizen who in the future will be able to generate benefits for society as a whole, being able to acquire the necessary experiences and evolutions fundamentally adapted to the context in which they find themselves.

The PIBID programme aims to: Encourage the training of teachers at higher education level for basic education, contributing to the enhancement of teaching; improving the quality of initial teacher training in degree courses, and simultaneously promoting dialogue between higher education and basic education; thus there is the possibility of inserting graduates in the daily life of schools in the public education network, providing them with greater contact related to the creation and participation in methodological, technological and teaching practice experiences of an innovative and interdisciplinary nature, seeking to overcome problems identified in the teaching-learning process. Encouraging public basic education schools, mobilising their teachers as shapers of future teachers and making them protagonists in the processes of initial training for the teaching profession. This will contribute to the articulation between theory and practice necessary for teacher training, raising the quality of academic actions in degree programmes (MEC, 2014, p.1).

To take part in PIBID, interested higher education institutions must first submit their respective projects to Capes (Coordination for the Improvement of Higher Education) in accordance with the selection notices published. With this in mind, both public and private institutions, whether for-profit or not-for-profit, can apply if they have a degree programme in common.

The institutions approved by Capes will receive grants and funding for the development of the project's activities related to the areas of research. Scholarship recipients are chosen through selections organised by each institution. The participants are: graduates, school supervisors, area coordinators, educational process management area coordinators and institutional coordinators. The scholarships are paid by Capes directly to the scholarship holders via bank credit. This is the start of the stages and activities that bring together the main issues related to a PIBID project.

The main aim of this work is to emphasise the importance of the PIBID project both for universities and their undergraduates, and for schools, their students and teachers, not forgetting to include the whole community, since the evolution of education is based on exercise and practice, with the aim of perfecting methods and contributing to the teaching-learning process, from the very beginning of academic life as a graduate student.

This organisation is based on the direct relationship with the dimensions and characteristics of the initiation to teaching provided for in Capes Ordinance No. 96/2013 in the following stages: Organisation and Preparation, Team Formation and/or Planning, Execution of Formative and Didactic Pedagogical Activities in Schools, Formative and Didactic Pedagogical Activities in the Field, Monitoring of the Project, Socialisation of Results, Presentation of Work at an Event in the Country, Presentation of Work at an Event Abroad, Publication of Digital Material and Accountability (CAPES, 2013, p. 21). It is hoped that PIBID will reduce dropout rates and increase demand for degree courses, recognise a new status for degree courses in the academic community and indicate an improvement in the IDEB in participating schools.

Therefore, the present discussion and observations highlight the importance of the Institutional Teaching Initiation Scholarship Programme (PIBID) for the training and practice of graduate teachers, thus pointing out educational measures and how to use teaching materials for better student conception. To this end, the aim is to identify the perception that these graduates have of the importance of the teacher training process based on the actions and activities experienced in the PIBID Programme.

2. Discussion

2.1 The importance of PIBID in teacher training

The choice to use Fonseca's studies has the function of grounding this work through her contribution to the discussion on teacher training and the reflections she brings to pedagogical practices.

In agreement with Fonseca (2002, p. 85-102), we start from the premise that the training process is not only built within and during the teacher training course. It is a daily construction, a dynamic process that takes place within a given socio-cultural reality. So being in training implies a personal, free and creative investment aimed at building a personal identity, which is at the same time a professional identity.

Consequently, practice is seen as an essential element in teacher training, but it needs to be seen not just as a field for applying theories, but as a source of possibilities, transforming lived experience into new knowledge and practices.

The programme offers scholarships for undergraduate students to carry out teaching activities in public basic education schools (both at municipal and state level), contributing to the process of integrating theory and practice, with the aim of bringing university and public education closer together. To ensure positive educational results, the scholarship holders are guided by area coordinators - university professors - and supervisors - teachers from the public schools where they work.

Since it was set up by Decree No. 7.219 of 24 June 2010, the programme has aimed to encourage teacher training at higher education level to work in basic education, helping to enhance the teaching profession by raising the quality of initial teacher training in degree courses, promoting integration between higher education and basic education. Inserting graduates into the daily life of public schools, providing them with opportunities to observe and participate in methodological experiments and innovative and interdisciplinary teaching practices that seek to overcome problems identified in the teaching-learning process. In addition, it contributes to the link between theory and practice that is necessary for teacher training, thus raising the quality of academic practice in teacher training courses.

Another important function attributed to the PIBID in relation to teacher training is pointed out by Gatti's research (2014, p. 9-155), which reports on the importance of qualification programmes in initial teacher training, mainly due to their positive results, which end up stimulating students to stay in the field and increasing the number of professionals available on the market.

Within this approach, the teacher and researcher Borges (2015), in her monograph, investigated and analysed the importance of PIBID in the practical training of mathematics graduates from the State University of Southwest Bahia (UESB) Vitória da Conquista campus. The results of her research provided a lot of relevant information to understand the importance of the project in the initial training of graduates. The research was based on a collection instrument and questionnaire.

The results show that the students' participation in the programme contributes to valuing the teaching profession and increasing the quality of initial teacher training in integration with public education. Through the students' reports, the researcher can conclude that PIBID plays a significant role in teacher training. It provides the individual in training with extremely important practical experiences, bringing together the theory provided by the university and pedagogical practice, progressing in their understanding of the content, their critical sense and their use of new teaching approaches and methods.

2.2. The Institutional Scholarship Programme for Teaching Initiation PIBID, pros and cons

Among the pros is the proposal for an innovative model with dynamic classes in which students reconnect with the school environment through workshops, debates, fairs, films and documentaries, organised by the PIBID fellows, always highlighting the community context, by using critical and more intimate forms about the society in which the student is inserted, thus, according to the authors Paschoal & Cavarsan, it should also be noted:

> Considering that the fellows had no experience of teaching, the first step was a visit to the project's partner school, which was very receptive, with

> the aim of recognising the physical conditions of the environment and a first contact with the supervising teachers. (PASCHOAL; CAVARSAN, 2014, p. 5)

In this way, the scholarship holders can exercise their academic learning by putting the theory they have learnt into practice and perfecting their educational practices in these moments of intersection that the PIBID provides.

On the other hand, contradicting the ideology mentioned above, there are conservative and traditional teachers who won't give up the old teaching model and who no longer contribute to the efficiency of today's learning. In this way, today's young people find themselves increasingly versed in the influence of the technological age, and thus increasingly connected to virtual media.

It's worth pointing out that not all teachers have the teaching resources they need to stimulate students and engage them in the content covered during lessons. As a result, there are insufficient methods to make the lesson essential and insightful, such as the use of outdated textbooks or content that is not worked on in a way that arouses the student's interest, leading to poor performance in the classroom.

Another negative aspect that jeopardises the efficiency of the project concerns the school itself not being completely open to the proposals presented, or even when the school completely refuses to take part in the project, attributing inefficiency, or just a lack of interest in giving up the old methodology, among other allegations.

We can also include in this context of challenges the fact that the project has few financial resources, requiring personal commitment from each participant in order to achieve its initial objectives as far as possible, without significant loss of inclusion and absorption of knowledge.

It's worth pointing out that at such a critical time as education is currently going through, with project cuts, interruptions, vetoes and a certain climate of uncertainty, the challenge becomes ever greater.

Thus, the project was created as a new way of re-establishing harmony and developing ideas between student and teacher, as well as making students more participative and inclusive in the classroom, as well as encouraging them to pursue a degree and a promising teaching future.

2.3 Practice and workshop developed by the Institutional Teaching Initiation Scholarship Programme (PIBID)

According to the theoretical and practical basis used in the classroom, soil is a natural environmental material that is very educational and easy to handle in a didactic way with the students, due to its assimilation and applicability as it is a natural work. In this way, when working with soil, the social aspect is brought into line with the reality of the human being in direct contact with the earth, enabling perspective-taking and physical relaxation.

With the material in practice, it can be seen that didactic material is a practical reference that guides new methods and influences the educational environment. So, when using soil as preparatory material for a lesson, it's worth first analysing the ways in which it can be used and for which target audience.

In order to carry out the soil paint workshop, it was necessary to use and prepare some materials for the purpose, such as brushes, containers, types of soil, glue and water. It is worth noting that it is also possible to find a variety of soil types on the earth's surface, demonstrating the diversity of colours that can be transformed into paints to work on in the classroom with students.

At the Dr Fernando Alexandre State School (2015), the didactics of soil paint were developed with the 4th grade students, which consisted of presenting the material that would be used, thus aiming to understand the students about the soil itself that covers the region in which it is inserted, thus giving purpose to the workshop.

In the final execution with the material, the students painted the canvases with soil paints taken from the school's own vegetables, thus obtaining an educational practice, both for cultural measures and in terms of art and teaching itself, demonstrating the effectiveness of the work done by the scholarship holders and the students within the schools.
in general. The project requires and relies on the interest of the students so that their performance in school, based on creative and motivating activities, contributes to their learning.

Conclusions

From the interventions of the work that PIBID has been developing over time, it has been possible to see the relevance of learning through the use of new pedagogical activities, new methodologies for teaching Geography. With this and good planning, good results can be achieved, with an adequate structure for developing quality work related to the school environment.

Currently, traditional lessons based on memorisation are considered to be tiring for the students, and as a result they lose interest, so it's necessary to get away from the textbook to do the practical work, entertaining the students more, which can generate greater interest. Therefore, the use of new teaching methods such as fieldwork, practical work, working with models, maps, satellite images and others can strengthen students' interest in lessons. By applying these alternative methods,

Photo 01: Making and practising with the material: Soil paint

Source: BORGES, M.A. (2015).

Photo 02: Completed material

Source: BORGES, M.A. (2015).

3. Final Results

PIBID is a project that allows contact with the school, helping in the academic training of undergraduates, so it is necessary to look for new practices that facilitate student learning, such as carrying out projects and educational practices.

the lesson becomes more dynamic and students participate, resulting in greater interaction and engagement with the subject and thus learning it. Nowadays, teaching methods can be reviewed and rethought so that they contribute to and facilitate the process of exchanging knowledge between teacher and student.

4. Thank you

We would like to thank CAPES for their support through the Institutional Teaching Initiation Scholarship Programme - PIBID/CAPES/FAI, which enabled us to carry out this work.

5. References

AMBROSETTI, N. B. et al. **Contributions of Pibid to initial teacher training.** Educação em Perspectiva, Viçosa, v. 4, n. 1, p. 151-174, Jan/jun. 2013.

BRAIT, Apud. **The teacher/student relationship in the teaching and learning process.** Electronic journal of the pedagogy course at the Jataí-UFG campus. V.8 N.1 Jan/Jul 2010.

BRAZIL. Decree No. 7.219, of 24 June 2010. Provides for the Institutional Teaching Initiation Scholarship Programme - PIBID and makes other provisions. **Federal Official Gazette**. Brasilia: Civil House of the Presidency of the Republic, 2010.

CAPES, **Coordination for the Improvement of Higher Education Personnel) Pibid Institutional Teaching Initiation Scholarship Programme.** Available at: <http//www.capes.gov.br/educação-básica/capespibid>. Accessed: 13 September 2017.

FONSECA, Selva Guimarães. **Knowledge of experience, life stories and teacher training**. In: CICILLINI, Graça Aparecida; NOGUEIRA, Sandra Vidal. School education: policies, knowledge and pedagogical practices. Uberlândia: EDUFU, 2002. p. 85-102.

CARLOS CHAGAS FOUNDATION. **An evaluation study of the Institutional Teaching Initiation Scholarship Programme (Pibid)** / Bernardete A. Gatti; Marli E. D. A. André; Laurizete Ferragut, researchers. Gatti; Marli E. D. A. André; Nelson A. S. Gimenes; Laurizete Ferragut, researchers. - São Paulo: FCC/SEP, 2014.

GADOTTI, M. Invitation to read Paulo Freire. São Paulo: Scipione, 1999.

GATTI, Bernardete Angelina; NUNES, Marina Muniz Rossa (Orgs.). **Teacher training for primary education**: *a* study of the curriculum for degrees in pedagogy, Portuguese language, maths and biological sciences. São Paulo: FCC, 2009. p. 9-155.

GOMES, A. M. A. et al. **Pedagogical knowledge and practice: an integration between theory and practice.** Educar, Curitiba, n. 28, p. 231-246, 2006. Editora UFPR.

GOMES, L. S. **The importance of PIBID in the training and teaching practice of maths graduates from the UESB Vitória da Conquista Campus.** Bahia. UESB, 2015.

HARVEY, D. **The postmodern condition**. São Paulo: Edições Loyola, 1993. Part III. p.219-276. MEC. **Ministry of Education MEC programmes for teacher training.** Available at: <http://portal.mec.gov.br/index.php?optio n=com_content&view=article&id =15944:programas-do-mec-voltados-a-formacao-de-professores>. Accessed on: 15 September 2017.

PASCHOAL, W.A.; CAVARSAN, E.C. **Impacts of the institutional teaching initiation scholarship project (pibid) on geography: experiences in critical citizen education. 2014, p.**
5. Available at: < https://www.uniso.br/publicacoes/a nais_eletronicos/2014/4_es_praticas_ed ucacionais/22.pdf >. Accessed on: 13 September 2017.

SANTOS, É. de S. PIBID - **INSTITUTIONAL PROGRAMME FOR INITIATION TO TEACHING: "A STUDY OF THE STATE OF THE ART"**. Electronic Journal "Diálogos Acadêmicos" JAN-JUL 2015

5. ETHNIC-RACIAL EDUCATION: Law 10.639/03 and the teaching of Geography.

Objectives

To deal with the implications of Law 10.639, which was passed in 2003, as well as the implications of how this law and its modifications are being applied. Based on authors who provide us with a better understanding of the subject, we will organise a piece of work in which the role of Geography in the application of the law is made clear, as well as how we can improve teaching and learning based on materials that provide us with better applicability of the law. Geography may be the discipline with the ability to relate the concepts that form opinions and associations, as well as to reduce or increase racial inequalities, which is why we have taken as our guide works by authors who have provided conceptions of how geography can be used in favour of Law No. 10.639.

Introduction:

Geography may be the discipline with the capacity to relate the concepts that form opinions and associations, as well as to reduce or increase racial inequalities through the formation of these concepts. For this reason, we have taken as our guide works by authors who have provided conceptions of how geography can be used in favour of Law No. 10.639. To this end, in this article we will look at Law No. 10.639 and its applicability, highlighting some historical facts and its dimensions with regard to the law and the teaching of geography. Throughout the article we will briefly discuss what the law is, its possible interpretations and changes, its developments and how the agents present in educational practices can effectively put it into practice. Law 10.639 of 2003 amended Law 9.394 of 1996, which established the guidelines and bases of national education. On 9 January 2003, after much struggle by the Black Rights Movements, the then president Luiz Inácio Lula da Silva sanctioned Law 10.639.

Signed into law in 2003, Law 10.However, at the time, many black movement activists knew that the law was only the beginning of a great battle, now being fought both inside and outside the school environment, since the law is only an instrument that obliges the teaching of Afro-Brazilian History and Culture, But a law doesn't change the thoughts built up by a racist society around the subject, so there was the challenge of bringing the school community together and synchronising the education system where teachers of subjects that don't cover the subject directly don't feel responsible for the propagation of racial inequalities.

> It encounters a school environment made up mostly of players who have not been prepared to build an anti-racist education, as well as inadequate teaching materials with aspects that support the reproduction of racism. It is in this environment that some teachers, parents, pedagogical coordinators, school management, as well as

> anti-racism activists, will fight over interpretations in the application of the law. But not only in the school environment: Academia, social movements and the state, all involved in the production and dissemination of reference materials, articulation and training of bodies and institutions that define and implement educational policies, production and dissemination of knowledge, among others, are arenas of dispute outside the school, but are also fundamental to the implementation of Law 10.639 (SANTOS, 2011, p. 7 and 8).

According to SANTOS, 2011, we can affirm that the law did come as an instrument to fight social inequalities, but we know that teachers have not been prepared for the effective application of this law, since teachers are the greatest agents of application, as long as they are really trained for this purpose.

Working Methodology

We have taken as our guide works by authors who have provided conceptions of how geography can be used in favour of Law No. 10.639. To this end, this article will look at Law No. 10.639 and its applicability, highlighting some historical facts and its dimensions in terms of the law and geography teaching. Throughout the article we will briefly address what the law is, its possible interpretations, its developments and how the agents present in educational practices can effectively put it into practice.For better development of the work and understanding of the framework regarding the applicability of Law 10.639 and the teaching of Geography, we used authors such as SANTOS, Renato Emerson; BOBBIO, Norberto; and GOMES, Norma Lino. These authors provided the best combination of references on the subject proposed in the work.

3. Results and Discussion

3.1 Law 10.639 and its amendments

In 1996, the then president of Brazil Fernando Henrique Cardoso approved Law° 9.394, of 20 December 1996, which establishes the guidelines and bases of national education, to include in the official curriculum of the Education Network the compulsory subject of "Afro-Brazilian History and Culture", and makes other provisions.

In 2003, President Luiz Inácio Lula da Silva sanctioned Law No. 10.639, which modifies Law No. 9.394, and its first paragraph makes its change clear where 1° The programme content referred to in the **caput of** this article will include the study of the History of Africa and Africans, the struggle of black people in Brazil, black Brazilian culture and black people in the formation of national society, rescuing the contribution of black people in the social, economic and political areas pertinent to the History of Brazil (Presidency of the Republic, Civil House, Sub-Cabinet for Legal Affairs, 2003). Law 11.645/2008 amends Law 9.394/1996, amended by Law 10.639/2003, and one of the main features of the amendment is the introduction of indigenous rights, as well as the compulsory teaching of Afro-Brazilian culture, making the teaching of indigenous matrices compulsory.

3.2 Theoretical Development

To better develop the work and understand the framework regarding the applicability of Law 10.639 and the teaching of Geography, we used authors such as SANTOS, Renato Emerson; BOBBIO, Norberto; and GOMES, Norma Lino. These authors provided the best combination of references on the subject proposed in the work. In his article A LEI 10.639 e o Ensino de Geografia: Construindo uma agenda de pesquisa - ação, published in 2011 by Revista Tamoios, the author takes us to reflect on how we got to the sanction of the law in 2003 and how it is being applied and how it should be applied, making us think about the paths that go beyond everyday school life and reach other spheres of the state that are also agents in the application of this law. In his considerations, the author brings us a concept that he classifies as dimensions, and in his view of how Geography and its learning is important in this process of applying the law, in these dimensions that are focused on the teaching of Geography, the role of the geography teacher in the intention of ending racial inequalities becomes clear. Based on this article, we have organised a table which brings together the dimensions classified by the author and their main characteristics as a concept also defined by the author.

Below is a demonstration table:

Table of dimensions and concepts in geography teaching, according to SANTOS, 2011.

DIMENSIONS	The association between racial groups and regions (geocultural) of origin.	Division dichotomised world (since Ratzel) between developed and underdeveloped countries.	From the spread of the monoculture of linear time(space), by the the way in which the role of technique as an evolutionary dimension is worked out.	The vision of the contemporary world as the overflow of economic, political, social, military and cultural processes from Europe.	Spreading a vision technical Cartesian world, e.g. through the how we teach cartography.
CONCEPTS	Give way to permanence of the idea of race as aregulator behaviours, values and social, economic and power relations;	And among them, the "developing" countries, which (a) reinforces the idea of a linear evolution in which the world's only future is to follow the path of the so-called "developed" countries, and (b) confers power in relations with the "developed" countries. social individuals and groupswhose historicity, they are referred to heritage and connection with these countries and peoples "developed" and therefore superior;	e.g. in the way in which the concept of landscape, through division between "natural landscapes" and "humanised landscapes", the latter always being (in evolutionary terms) the expression of technological advances ona terrestrial materiality. In this way, landscapes that are the result of simultaneous experiences appear to be landscapes of the past , landscapes of the present landscapes of the future;	This comes out very strongly in the way we teach aboutothers continents, whose historical references and spatial periodisation and regionalisation always appear as a direct result of the processes Eurocentred interests, as if there were no protagonist in them;	Which are, in other words, limited by this rationality to Western scientific ways of seeing the world, of expressing references to space, of time and social existence. This way of working teaches School cartography gives official maps the character of an expression of truth that is a powerful instrument of power through the production of non-existence of social groups, conflicts, knowledge, experiences and form s of behaviour. relationship with the world.

source: Law 10.639 and Geography Teaching: Building a
research-action agenda .ÓRG: VALENTIM, R. 2018.

According to the table, we have seen that the role of Geography teaching as an instrument of social transformation with regard to Law No. 10.639 is present in the school environment and outside of it, since through classifications originating in geographical concepts they are in vogue both within school spaces, as well as in graduate training spaces that will also help in the construction of ideas that permeate this principle of racial classification, according to SANTOS, 2011. Geography is therefore, in a very subliminal way, at the base of the construction of ideas, relationships and behaviours based on the principle of racial classification. The worldview that Geography constructs underpins racial identities. Geography teaching is one of the main vehicles (SANTOS, 2011, p. 11).

From this concept of vehicle that SANTOS brings us, he explains to us the dimensions that the teaching of Geography can allow a vision of the world in which individuals are placed in roles of prominence or not. The prominent role is given to individuals who refer to the heritage of peoples considered to be developed, thus playing a crucial role in existing race relations inside and outside the school environment, making racist actions and relations seen as normal and supporting the naturalisation of social inequalities, as has been widespread for centuries in Brazilian and even world education. These five dimensions described by SANTOS show us how the teaching of Geography can be decisive in the process of making Law 10.639 effective. For better elucidation, we will present these five dimensions according to the author and, for better understanding, we will present a table with the author's words so as not to generate any kind of interpretation that is not in line with the author's idea and conceptualisation. From another point of view, but without detracting from the racial classifications, GOMES, 2005, brings us the vision used in the racial classification of yesteryear, which largely contaminated the construction of knowledge from the early years of education, where racial classification was made without questioning why each one occupies this or that role in which it is inserted in the didactic materials used in the construction of teaching, and which is largely responsible for maintaining racial inequality.

> It is important to emphasise that, in this sense, races are understood as social, political and cultural constructions produced in the context of power relations throughout the historical process. They are by no means a given of nature.2 It is in culture and social life that we learn to see races. This means that we learn to see people as black and white and, consequently, to classify them and perceive their differences in social contact, in the way we are educated and socialised to the point where these so-called differences are introjected into our way of being and seeing others, into our subjectivity, into wider social relations. In culture and society, we learn to perceive differences, to compare, to categorise. If things were just on that level, we wouldn't have so many complications. The problem is that, in this same context, we learn to hierarchise social, racial and gender classifications, among others. In other words, we also learn to treat differences unequally.(Gomes, N. 2011)

3.2 Instruments for implementing Law 10.639

Based on the study carried out by SANTOS 2011, in which he presents us with the five strands used by him and by teachers who take part in a group that aims to help in the fight against racial inequalities by applying Law n°10.639, we will present the main aspects of these strands in order to make it clear how it is possible to teach geography that does not serve to support the racial inequalities that are already present in everyday school life and beyond, but rather transforms them through the teaching of geography. With this change, the aim is to serve as a basis for the construction of thoughts in which social inequality is seen without masks and is shown in a way that was never made explicit in the school environment until the enactment of Law No. 10.639.

The first aspect is the insertion and revision of programme content in the Geography curriculum, which in a nutshell is the insertion of content that should have been in the current curricula but wasn't, and it's not enough to insert content on the History and Culture of Afro-descendants or Africa. In the second strand we have the Review of Pedagogical Practices, Materials and Methods, which implies a review of the most widely used teaching material, the textbook. From this review we will have a different look at the issue and how it is dealt with in the textbooks, as well as how to apply what is in the textbooks in order to curb the racism present in the textbooks and that goes beyond the pages of the textbook. In this aspect we would have to critically analyse how much black people are shown in the textbooks, as well as the way they are shown in the textbook, in short, analyse the structure of the textbook and how it is proposing the law. The third aspect is the Management of Race Relations in Everyday School Life. In order to enforce the law, which in theory introduces and modifies race relations through the aspects mentioned above, it is also necessary to transform everyday school life, which is already a racist environment, and most of whose content ratifies and justifies racist actions in the classroom or school space. It is believed that by meeting the content demands set out in Law 10.639, these relationships will be positively impacted by educating for racial equality (SANTOS, 2011. p 18). In the fourth strand, power relations come into play in the construction of the curriculum practised at school. In this strand, we see that in order for the law to be applied, everyone in the school environment must have the same intention of enforcing the law, which is why this fourth strand is so important. It is through these multiple roles played in everyday school life that dialogues are built about how this law can be applied, and according to the author these dialogues are made up of interests that sometimes clash and sometimes complement each other.

> This places the treatment of Law 10.639 as an object, as an agenda for negotiation in everyday school life. These negotiations involve multiple actors: teachers, management, pedagogical coordination, support staff, other spheres of institutional coordination at network level, etc. Our observations show us how crucial it is for the implementation of the Law to understand how it is influenced by power relations that interfere with the construction of the curriculum that is actually practised. Official arenas (class councils, pedagogical meetings, disciplinary forums, etc.) and hidden arenas (breaks in the teachers' room, informal meetings in the corridor, the canteen and other spaces); official agendas (Pedagogical Political Project, Project Activities, cross-cutting themes to be worked on by the school) and hidden agendas (power struggles within the school or within the network, the formation of groups based on personal, ideological or other affinities, etc.).); all of these dimensions influence the implementation of the law, and their decisive nature has made them the object of investigation within the scope of the research.(SANTOS,2011.p 19)

4. Final considerations

Based on the study, we concluded that Law 10.639 is here to serve as an instrument against racial inequality and that everyone needs to be committed to enforcing the law. Sanctioned in 2003, little is heard about the application of this law, which if applied from the beginning of its sanction would already have at least satisfactory results, given that its applicability depends on teachers, but also on everyone who makes up the school environment, from students to the MEC. What we have seen from this article is precisely the opposite of efficiency when it comes to applying the law and its positive impact on society. There is no interaction between the disciplines for the law to work, and it has become common sense even among teachers that the law is not for all subjects, but only for subjects that are directly linked to the history and culture of Afro-Brazilians and Africa itself, becoming the sole responsibility of teachers directly linked to these themes proposed by the law, without having the notion that the disciplines talk to each other and that one is intimately linked to the other and that they all make up elements of the individual's personal and intellectual formation, notwithstanding the teaching of geography, which is capable of transcending other disciplines.

In the teaching of Geography, hopes are renewed, given that the curriculum itself is very flexible in terms of changes, not just in the textbook, but also in the teacher themselves, who can complement the content without precisely changing the textbook, making the teacher an essential instrument in the construction of anti-racist thoughts. From geographical concepts that form racist concepts, we can transform them so that they are used in a different way and form concepts that refer to racial equality, but these teachers need training for something that looks easy on paper, but in reality takes years and years of hard work.

In conclusion, we consider Law 10.639 to be a good instrument against racial inequality, but this instrument needs to be polished so that it is used in the right way and can bring positive repercussions to society, in a general context, the Law is beneficial, but we cannot let this law be seen as the same as the PPPs, which exist, but have become mere drawer instruments in most public schools and are not followed in the way they should be.

References

SANTOS, Renato Emerson dos. Rediscussing Geography Teaching: Themes from Law 10.639. Rio de Janeiro: CEAP, 2009.

SANTOS,Renato Emerson dos.A Lei 10.639 e o Ensino de Geografia: Construindo uma agenda de pesquisa-ação.Revista Tamoios. Year VII. N° 1, 2011 - ISSN 19804490, 2011.

CAMPOS, Andrelino. From the Quilombo to the Favela: The production of "criminalised space" in Rio de Janeiro. Rio de Janeiro: Bertrand Brasil, 2005

http://www.planalto.gov.br/ccivil_03/Leis/2003/L10.639.htm(accessed on 06/07/2018)

BOBBIO, Norberto et al. Dictionary of politics. Brasília: Ed. Universidade de Brasília, 1992.

CASHMORE, Ellis. Dictionary of ethnic and racial relations. São Paulo: Selo Negro, 2000.

GOMES, Nilma Lino. Some terms and concepts present in the debate on race relations in Brazil: a brief discussion. Anti-racist education: paths opened up by Federal Law No. 10.639/03. Brasília:

MEC/SECAD, 2005. p. 39-62.

I want morebooks!

Buy your books fast and straightforward online - at one of world's fastest growing online book stores! Environmentally sound due to Print-on-Demand technologies.

Buy your books online at
www.morebooks.shop

Kaufen Sie Ihre Bücher schnell und unkompliziert online – auf einer der am schnellsten wachsenden Buchhandelsplattformen weltweit! Dank Print-On-Demand umwelt- und ressourcenschonend produziert.

Bücher schneller online kaufen
www.morebooks.shop

MIX
Papier aus verantwortungsvollen Quellen
Paper from responsible sources
FSC® C105338